W9-CEW-102

BIG MACHINES

Written by Melanie Davis Jones
Illustrated by Doreen Gay-Kassel

Children's Press®
A Division of Scholastic Inc.
New York • Toronto • London • Auckland • Sydney
Mexico City • New Delhi • Hong Kong
Danbury, Connecticut

To my Daddy, who has driven many big machines, and to my boys who love to watch them

—M.D.J.

To Lewis, who is my favorite driver

—D.G.-K.

Reading Consultants

Linda Cornwell
Literacy Specialist

Katharine A. Kane
Education Consultant
(Retired, San Diego County Office of Education
and San Diego State University)

Library of Congress Cataloging-in-Publication Data

Jones, Melanie Davis.
Big machines / written by Melanie Davis Jones ; illustrated by Doreen Gay-Kassel.
p. cm. — (Rookie readers)
Summary: A simple introduction to big machines—tractors, backhoes, pavers, and cranes—and how they work.
ISBN 0-516-22845-5 (lib. bdg.) 0-516-27829-0 (pbk.)
1. Earthmoving machinery—Juvenile literature. [1. Earthmoving machinery.] I. Gay-Kassel, Doreen, ill. II. Title. III. Rookie reader.
TA725 .J66 2003
629.225—dc21 2002008780

4 5 6 7 8 9 10 R 12 11 10 09 08 07 06 05 04 03

A tractor pulls.

A tractor plows.

A tractor hauls hay
for the cows.

A backhoe digs.
A backhoe dumps.

A backhoe scoops up
big tree stumps.

A paver smoothes.
A paver rolls.

A paver fills in
great big holes.

A tall crane lifts.
A tall crane loads.

A tall crane helps
to build the roads.

Big machines are moving dirt.

Big machines are hard at work.

Word List (39 words)

a	great	pulls
are	hard	roads
at	hauls	rolls
backhoe	hay	scoops
big	helps	smoothes
build	holes	stumps
cows	in	tall
crane	lifts	the
digs	loads	to
dirt	machines	tractor
dumps	moving	tree
fills	paver	up
for	plows	work

About the Author

Melanie Davis Jones grew up on a farm in Georgia. While living on the farm, she saw her father drive tractors, combines, backhoes, and many other big machines. Mrs. Jones is now a teacher and children's author. She and her husband also build houses. While they build the houses, her sons love to play in the dirt and watch big machines at work. That is what inspired this book.

About the Illustrator

Doreen Gay-Kassel works with clay all day long in her studio in a wonderful old house that she shares with her talented sons and husband and their offbeat Jack Russell terrier, Rosie.